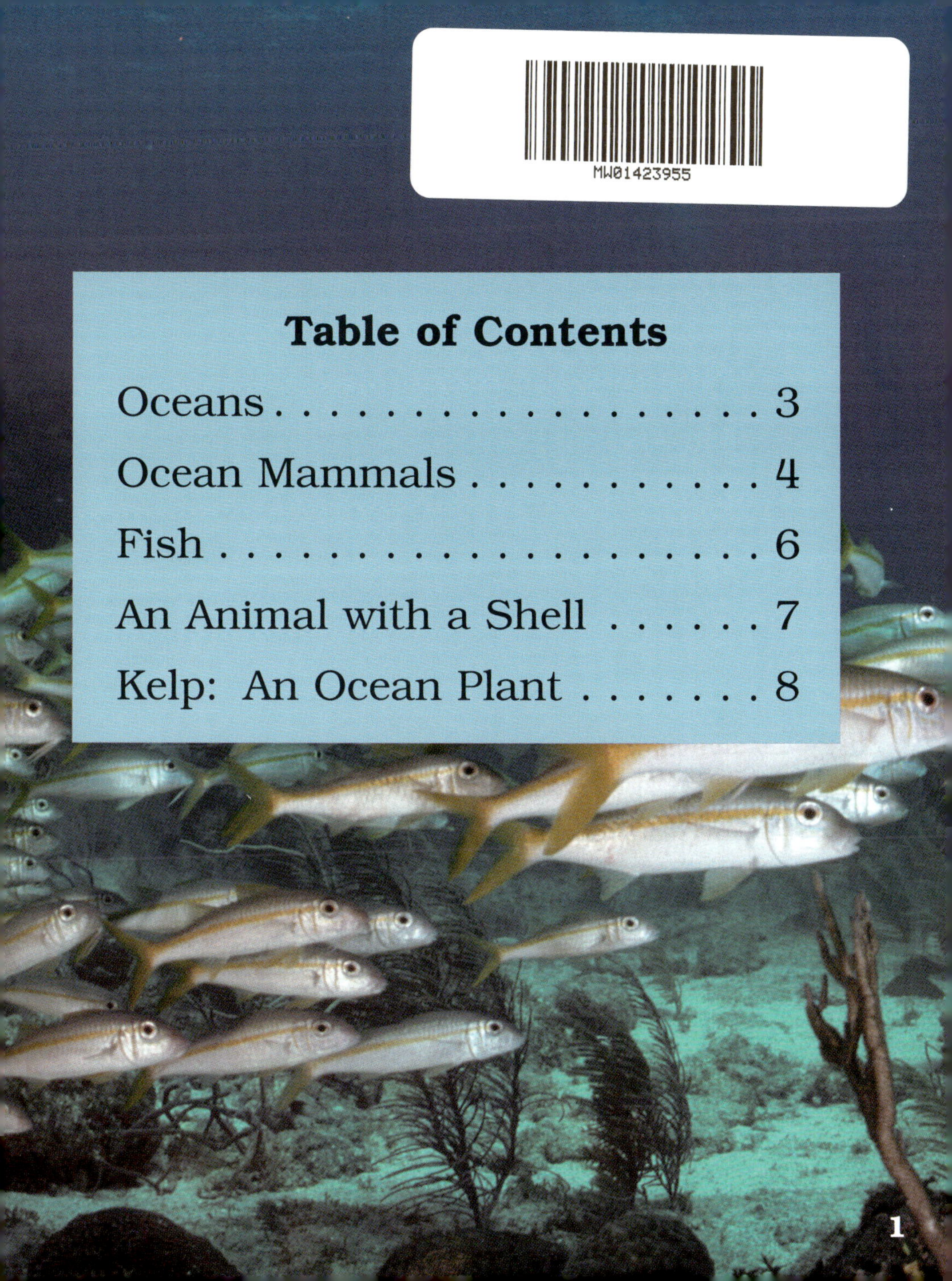

Table of Contents

Oceans 3
Ocean Mammals 4
Fish 6
An Animal with a Shell 7
Kelp: An Ocean Plant 8

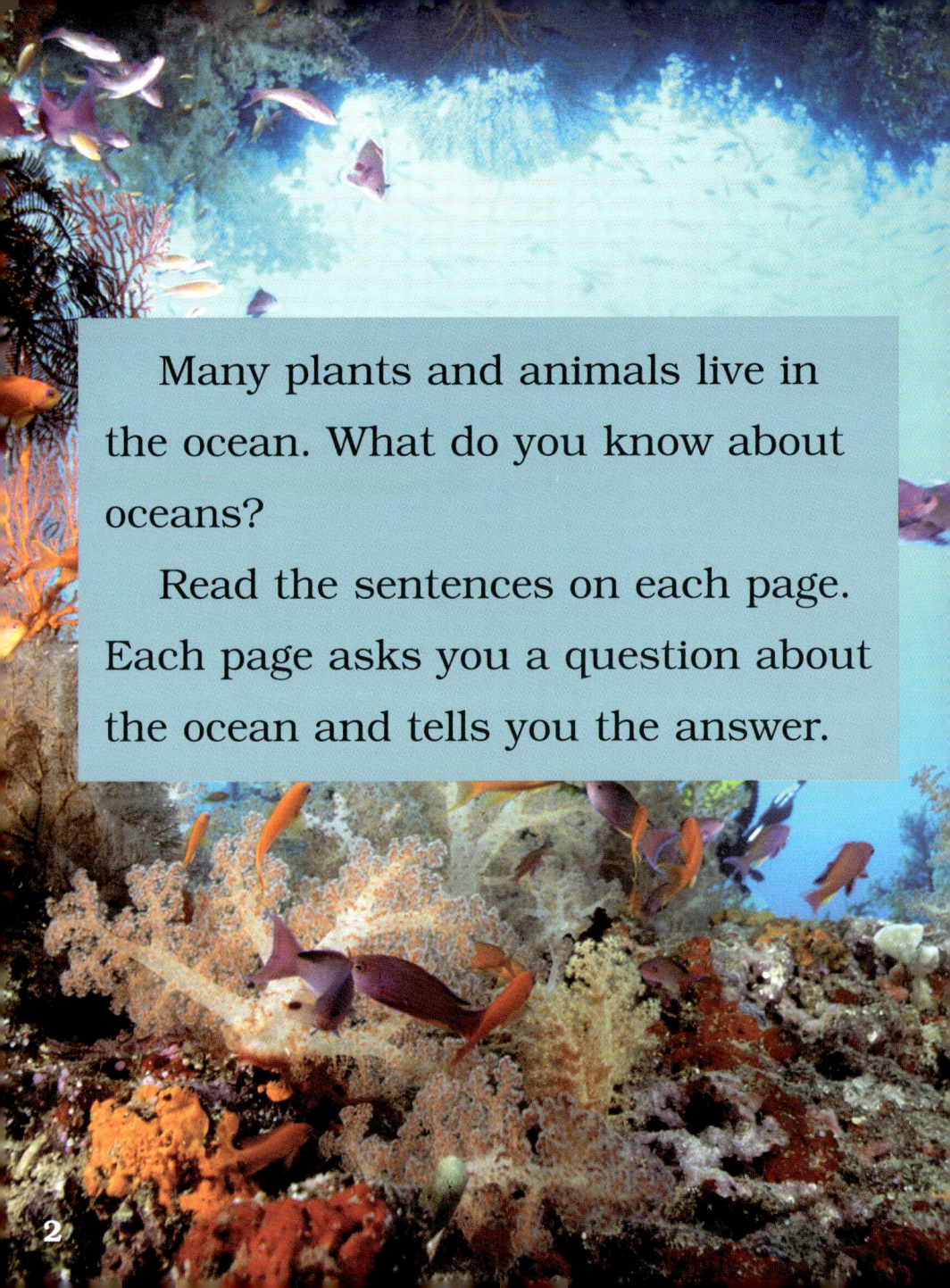

Many plants and animals live in the ocean. What do you know about oceans?

Read the sentences on each page. Each page asks you a question about the ocean and tells you the answer.

Oceans

Question: Can you guess what the ocean is like?

   a. salty

   b. sweet

Answer:

a. The ocean is salty.

Ocean Mammals

Question: Seals like the ocean. What is a young seal called?

    a. kitten

    b. pup

Answer:

b. A young seal is a pup.

Question: In what place do polar bears swim?

    a. Arctic Ocean

    b. Atlantic Ocean

Answer:

a. Polar bears swim in the icy cold Arctic Ocean.

Fish

Question: Why does the pilot fish swim under a shark?

    a. to play as friends

    b. to hunt for food

Answer:

    b. The pilot fish eats the scraps from the shark's meals.

An Animal with a Shell

Question: Are there enough sea turtles in the oceans?

    a. yes

    b. no

Answer:

b. No, most sea turtles are gone. Now there are laws against catching them.

Kelp: An Ocean Plant

Question: Why do animals need kelp?

   a. as a home

   b. as a place to rest again and again

Answer:

a. and **b.** Many sea animals live, hunt, and rest in kelp beds.